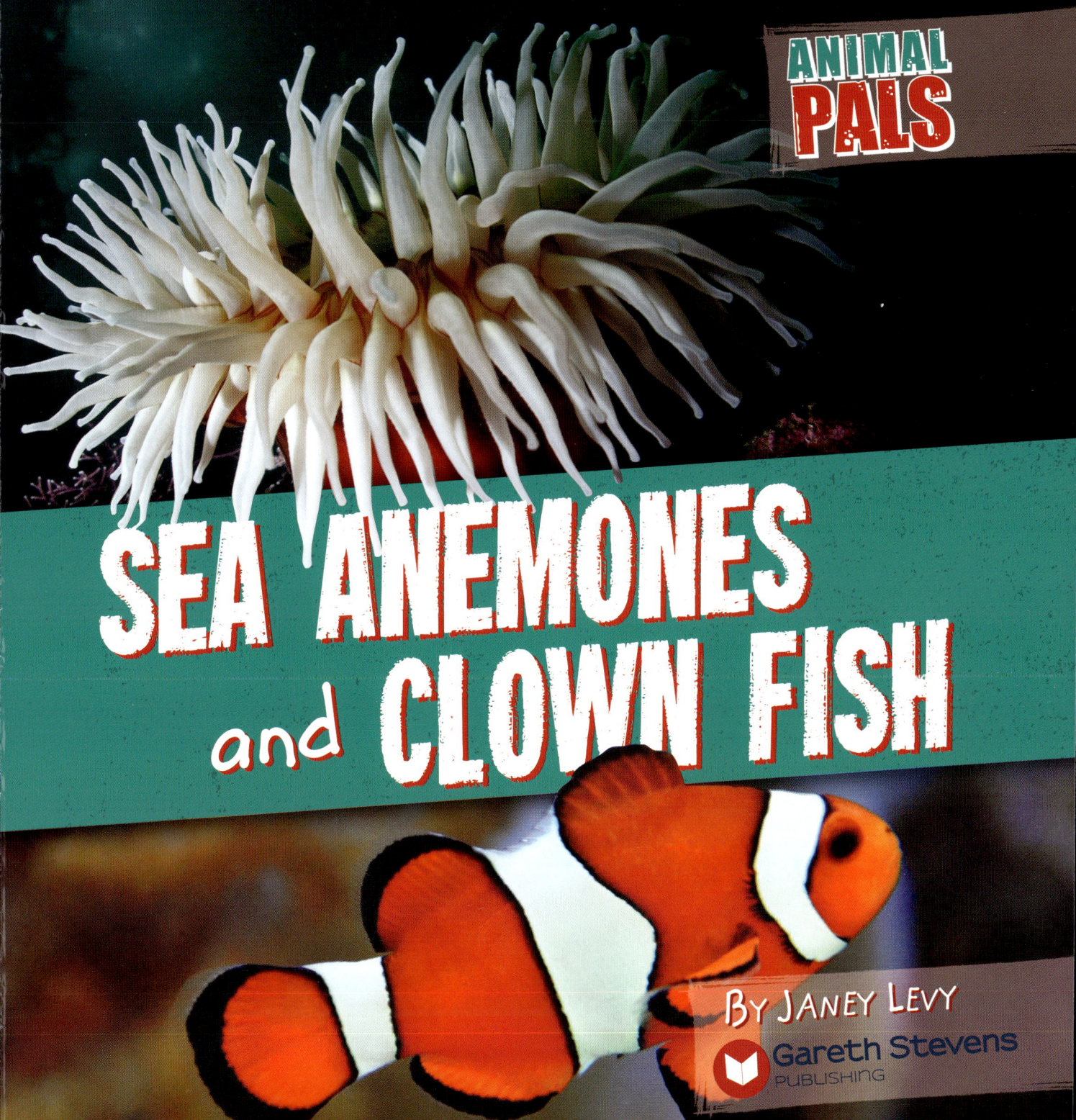

Animal Pals

SEA ANEMONES and CLOWN FISH

By Janey Levy

Gareth Stevens
PUBLISHING

Please visit our website, www.garethstevens.com. For a free color catalog of all our high-quality books, call toll free 1-800-542-2595 or fax 1-877-542-2596.

Library of Congress Cataloging-in-Publication Data

Names: Levy, Janey, author.
Title: Sea anemones and clown fish / Janey Levy.
Description: New York : Gareth Stevens Publishing, [2022] | Series: Animal pals | Includes index.
Identifiers: LCCN 2020035842 (print) | LCCN 2020035843 (ebook) | ISBN 9781538266892 (library binding) | ISBN 9781538266878 (paperback) | ISBN 9781538266885 (set) | ISBN 9781538266908 (ebook)
Subjects: LCSH: Sea anemones–Juvenile literature. | Anemonefishes–Juvenile literature.
Classification: LCC QL377.C7 L65 2022 (print) | LCC QL377.C7 (ebook) | DDC 593.6–dc23
LC record available at https://lccn.loc.gov/2020035842
LC ebook record available at https://lccn.loc.gov/2020035843

First Edition

Published in 2022 by
Gareth Stevens Publishing
29 E. 21st Street
New York, NY 10010

Copyright © 2022 Gareth Stevens Publishing

Designer: Andrea Davison-Bartolotta
Editor: Monika Davies

Photo credits: Cover (top), p. 1 (top) Ethan Daniels/Shutterstock.com; cover (bottom), p. 1 (bottom) Alisha Duponchel/EyeEm/Getty Images; p. 5 Federica Grassi/Moment/Getty Images; p. 7 (background) Kikinusska/Shutterstock.com; p. 7 (main) ArtMari/Shutterstock.com; p. 8 Fotosearch/Getty Images; p. 9 (top left) Tharapong Talubnak/Shutterstock.com; p. 9 (top right) Andrea Izzotti/Shutterstock.com; p. 9 (bottom left) Natalia Fedori/Shutterstock.com; p. 9 (bottom right) David A Litman/Shutterstock.com; p. 10 Giordano Cipriani/The Image Bank/Getty Images; p. 11 (top left) Christian Gerhardt/EyeEm/Getty Images; p. 11 (top right) Brandi Mueller/Moment/Getty Images; p. 11 (bottom left) Arunee Rodloy/Shutterstock.com; p. 11 (bottom right) Norman Lopez/EyeEm/Getty Images; p. 12 In Green/Shutterstock.com; p. 13 Dudarev Mikhail/Shutterstock.com; p. 15 Lotus41/Moment/Getty Images; p. 17 MariMore/Shutterstock.com; p. 19 TatianaMironenko/iStock/Getty Images Plus/Getty Images; p. 20 Paul Cowell/EyeEm/Getty Images; p. 21 Razvan Ciuca/Moment/Getty Images.

All rights reserved. No part of this book may be reproduced in any form without permission in writing from the publisher, except by a reviewer.

Printed in the United States of America

CPSIA compliance information: Batch #CSGS22: For further information contact Gareth Stevens, New York, New York at 1-800-542-2595.

CONTENTS

Reef Roommates 4

Salute the Sea Anemone 6

Amazing Anemones. 8

Colorful Clown Fish 10

Cozy Coral Reef Home 12

Why Live in a Sea Anemone? 14

Staying Safe Among Terrible Tentacles 16

How Clown Fish Aid Sea Anemones 18

Breathing Easy. 20

Glossary. 22

For More Information 23

Index 24

Words in the glossary appear in **bold** type the first time they are used in the text.

REEF ROOMMATES

Have you seen the movie *Finding Nemo*? The star of the movie is a bright orange clown fish with white stripes. Clown fish live in coral reefs and build their homes in creatures called sea anemones.

Sea anemones have harmful stinging **tentacles**. So, how are clown fish able to live among those tentacles? And why do they choose sea anemones as their home? Inside this book, you'll find the answers to these questions and also learn lots more about clown fish and sea anemones.

FACT FINDER!

Clown fish and sea anemones have a mutualistic **relationship**. This is a relationship between two different kinds of animals that benefits both of them.

Clown fish and sea anemones help each other survive in many ways.

SALUTE THE SEA ANEMONE

Colorful sea anemones are sometimes called flowers of the sea. In fact, they're named after garden flowers called anemones. But don't be fooled. Sea anemones may look like flowers, but they are actually animals.

Sea anemones have a tube-shaped body and a sticky, disk-shaped foot that holds them in place. A mass of tentacles surrounds their central mouth. Their tentacles carry stinging cells called nematocysts. These have poison to **paralyze** their **prey**. Sea anemones can then use their tentacles to move their meal to their mouth.

FACT FINDER!
Sea anemones poop through their mouth. Yuck!

The Parts of a Sea Anemone

- mouth
- tentacles
- body
- foot

Sea anemones don't have a brain!

7

AMAZING ANEMONES

There are around 1,000 species, or kinds, of sea anemones across the world. Sea anemones can be tiny or huge. They may be about 0.5 inch (1.3 cm) across to 6 feet (1.8 m) across.

Sea anemones can live to be up to 80 years old. While most sea anemones live near the ocean's surface in warm waters, some live over 32,800 feet (10,000 m) under the surface of the water. And they can be fighters! Colonies may fight one another to **protect** their territory.

Only 10 species of sea anemones are known to host clown fish. However, scientists believe there are probably more species that pair up with clown fish.

COLORFUL CLOWN FISH

The most famous species of clown fish is the ocellaris clown fish, or the common clown fish. It has a bright orange body with three broad white bands that have narrow black borders. But this isn't the only species of clown fish. There are around 30 species of clown fish—and they come in all kinds of colors!

Clown fish only grow to about 4.3 inches (11 cm) long. Clown fish are also omnivores, which means they eat both plants and animals.

FACT FINDER

Clown fish can be both male and female during their lives. They grow up as males first.

Surprisingly, clown fish don't swim very well!

COZY CORAL REEF HOME

Although sea anemones grow in oceans worldwide, clown fish live only in a few kinds. They live in sea anemones found on coral reefs in certain warm waters.

A coral reef is an underwater hill made up of many different species of corals. Corals are tiny ocean animals with hollow bodies and an inner or outer **skeleton**. A coral reef is a huge construction. The top of a reef has living corals, while the lower part is made of skeletons of corals that died long ago.

FACT FINDER
Some coral reefs that exist today are 10,000 years old!

Clown fish and sea anemones often pair up in coral reefs found in the Indian Ocean, the Red Sea, and the western part of the Pacific Ocean.

WHY LIVE IN A SEA ANEMONE?

Sea anemones have dangerous tentacles. They use their tentacles to capture their prey—which includes small fish. So, why would clown fish want to live in sea anemones?

Clown fish benefit greatly from living among sea anemone tentacles. Those deadly tentacles protect clown fish from predators that would eat them. Clown fish also get to eat bits of food left over from the sea anemone's meals. But how do clown fish manage to live among the poisonous tentacles?

FACT FINDER

Clown fish eat more than just the scraps from the sea anemone's meals. They eat tiny plants and animals floating in the water as well as **algae**.

Clown fish also eat dead parts of the sea anemone.

STAYING SAFE AMONG TERRIBLE TENTACLES

Clown fish can live within sea anemones without being stung. Their secret to staying safe is a layer of mucus!

Mucus is a thick slime produced by the bodies of many animals. Some clown fish are born with a coating of mucus on their bodies. Others must build up a coat of mucus before living in a sea anemone. These fish will brush up against an anemone's tentacles again and again. After a while, the clown fish builds up enough mucus to withstand the tentacles' sting.

FACT FINDER

Scientists think the coat of mucus on a clown fish grows to include mucus from the sea anemone. This mix of mucus creates a safe cover for the clown fish.

All fish have a mucus coating. But clown fish have a mucus coating that is three to four times as thick as the one other fish have!

HOW CLOWN FISH AID SEA ANEMONES

Clown fish have a mutualistic relationship with sea anemones. That means sea anemones also benefit from this friendship.

Clown fish are a great help to sea anemones. They help clean and remove **parasites** from the sea anemone. Clown fish also chase away predators such as butterfly fish. Without this protection, predators would eat the sea anemone's tentacles. Sea anemones also get important nutrients from clown fish waste. Nutrients are what a living thing needs to grow and stay alive.

This is a butterfly fish. Clown fish protect sea anemones by driving these fish away.

19

BREATHING EASY

Just like you, sea anemones and clown fish need **oxygen** to live. These two creatures breathe in oxygen from the water. But sea anemones, which stay in one place, have trouble getting enough oxygen. They can wave their tentacles to move water and bring more oxygen to them. But their weak tentacles can't move much water.

Scientists have discovered that clown fish twist and squirm a lot among the tentacles of a sea anemone. This moves more water around, and the sea anemones get more oxygen!

Clown fish and sea anemones have a friendship that helps them both. That's a mutualistic relationship in action!

GLOSSARY

algae: plantlike living things that are mostly found in water

oxygen: a colorless, odorless gas that many animals, including people, need to breathe

paralyze: to make unable to move

parasite: a living thing that lives in, on, or with another living thing and often harms it

prey: an animal that is hunted by other animals as food

protect: to keep safe

relationship: a connection between two living things

skeleton: the strong frame that supports an animal's body

tentacle: a long, thin body part that sticks out from an animal's head or mouth

FOR MORE INFORMATION

Books

Cunningham, Kevin. *Clownfish and Sea Anemones*. Ann Arbor, MI: Cherry Lake Publishing, 2017.

Rake, Jody S. *Sea Anemones*. North Mankato, MN: Capstone Press, 2017.

Schuetz, Kari. *Clownfish and Sea Anemones*. Minneapolis, MN: Bellwether Media, 2019.

Websites

Are Sea Anemones Plants or Animals?
www.wonderopolis.org/wonder/are-sea-anemones-plants-or-animals
Discover more about sea anemones at this website.

Clown Anemonefish
kids.nationalgeographic.com/animals/fish/clown-anemonefish/
Learn more about clown fish and their partnership with sea anemones here.

Common Clownfish
oceana.org/marine-life/ocean-fishes/common-clownfish
Find out more about clown fish here and watch a great video.

Publisher's note to educators and parents: Our editors have carefully reviewed these websites to ensure that they are suitable for students. Many websites change frequently, however, and we cannot guarantee that a site's future contents will continue to meet our high standards of quality and educational value. Be advised that students should be closely supervised whenever they access the internet.

INDEX

algae, 14

brain, 7

butterfly fish, 18, 19

colonies, 8

coral reefs, 4, 12, 13

corals, 12

flowers, 6

foot, 6, 7

Indian Ocean, 13

mouth, 6, 7

mucus, 16, 17

mutualistic relationship, 4, 18, 21

nematocysts, 6

nutrients, 18

ocellaris/common clown fish, 10

omnivores, 10

oxygen, 20

Pacific Ocean, 13

parasites, 18

predators, 14, 18

prey, 6, 14

Red Sea, 13

skeleton, 12

species, 8, 9, 10

tentacles, 4, 6, 7, 14, 16, 18, 20